The San Diego-La Jolla Underwater Park Ecological Reserve

A Field Guide

Judith Lea Garfield

Picaro Publishing - San Diego, California

Published by Picaro Publishing, San Diego 92169

First edition published in 1994.
Printed in Singapore.

99 98 97 96 95 94 6 5 4 3 2 1

Photo credits: All photos by the author except as noted;
Fred Fischer, pp.26, 53(Female); Richard Herrmann, p.32;
Steve Gardner, pp.34, 50(Adults); Phil Colla, pp.52, 53(Male), 58

Garfield, Judith Lea.
The San Diego-La Jolla Underwater Park Ecological Reserve, A Field Guide.

Library of Congress Catalog Card Number 94-66147
ISBN 0-9640724-6-7

To order additional books, send $9.95 plus $3.00 for shipping and handling to:
Picaro Publishing, Box 99304, San Diego, California 92169

Front Cover: La Jolla Cove, La Jolla, California

To Bob, Sasha, and Sydney
for their plenary love and support.

To my parents
for raising me to believe in all
of life's possibilities.

To everyone
who appreciates the subtle ways that
nature enriches our lives.

Foreword

This useful field guide is a must for novice and experienced snorkelers or scuba divers who want to learn about the marine life in La Jolla Cove without taking a marine biology course. It contains a wealth of information that will be useful to readers both before and after a dive. The short history of the Underwater Park and the development of the La Jolla Cove area are of particular interest. Judith Garfield has managed to present the Cove and its ambiance in the mindset of the "locals," which is not easily absorbed through a casual visit to the area.

Detailing the marine life found within the confines of the entire San Diego-La Jolla Underwater Park would be an enormous project. Instead, Judith has focused on the most widely explored part of the Ecological Reserve, the waters directly off La Jolla Cove. The Reserve is one of the best-known dive sites in Southern California, and La Jolla Cove is where the majority of people enter the water. This field guide provides the information necessary to plan a rewarding dive.

I particularly like Judith's suggestions for maximizing the sightings made in the course of a fun dive; few field guides teach how to be on the lookout for opportune creatures that you might encounter during a dive. Judith shares her insights, gained from many years of diving and underwater photography, in the parlance of the casual diver.

In the cool, temperate waters off La Jolla's coast, the number of species encountered is larger than most divers can identify. This guide includes most of the common species of marine plants and invertebrates, as well as the vertebrate world of fishes, birds, and mammals. The guide's format includes a succinct description of each species' appearance, size range, typical locale, and feeding habits, along with a paragraph of enchanting and informative tidbits to enlighten the reader. Color photographs depict each species in its natural habitat.

Judith's plea for the protection of this environment and its marine life is timely, as the Ecological Reserve has suffered from poachers and heavy traffic from many environmentally irresponsible visitors. Protection of the oceans must be spearheaded by the novice naturalist who understands that degradation can occur without care and gentle handling of the marine flora and fauna.

Well organized, detailed, and comprehensive, this guide gives all visitors the opportunity to identify and learn about the marine life found within the Ecological Reserve.

Dr. Bert Kobayashi
Emeritus Supervisor of Physical Education (Specialist in Scuba)
University of California, San Diego

Contents

Acknowledgments

A debt of gratitude is owed to the following: Rick Kern, dive buddy extraordinaire; Martha Blane, greatest friend and supporter; Annanda Stevenson, for finding my chi; Lee Peterson, for years of technical advice; Mary Rose, for cheerfully repairing my equipment; and Peter Brueggeman, for his wealth of library knowledge and enthusiastic sharing of ideas.

⟡

Special recognition goes to Dr. Bert Kobayashi for his long-term support and involvement in the Underwater Park. In 1978, he carefully designed and implemented a comprehensive survey of the Ecological Reserve. The survey, funded by the State Water Resources Control Board, detailed the topography and marine life of the area. The resulting published work remains an important historical document and an invaluable reference for the community.

⟡

For those assisting in the preparation of this book, many thanks go to the following: Diane Gage, great teacher and content editor; Bert Kobayashi, for writing the Foreword and overall technical editing; JoAnn Padgett, for outstanding copy editing; Pat Schaelchlin, from the La Jolla Historical Society, for sharing her knowledge of early La Jolla; Phil Colla, Fred (Red) Fischer, Steve Gardner, and Rich Herrmann, for helping me complete my photographic list; Chris Simpson, for developing the inside front cover map; and Blaise Nauyokas, for a knockout layout.

Introduction

The Ecological Reserve, a part of the San Diego-La Jolla Underwater Park, was created to protect marine life and its habitats from human intervention and is the only reserve of its kind in San Diego. Laws forbid anyone from removing a single plant, animal, or inanimate object from the Reserve.

La Jolla Cove, the beachgoers' entrance to the Reserve, is a part of this underwater sanctuary. "The Cove," as it is called by locals, is considered the most prime piece of underwater real estate in the region because of its beauty and proximity to three diverse habitats: rocky reef, sandy flat, and kelp bed. Each habitat shelters marine life unique to its own niche.

The Reserve, a long-time destination for tidepoolers, swimmers, snorkelers, scuba divers, visitors, and locals, can be found along the rugged coastline outlining La Jolla. An outcropping of rocks around the Cove frames a pool of shallow, smooth water. Numerous species of flora and fauna are easily observed in this part of the Reserve and in the waters surrounding the Cove. Depths average less than 10 feet around the rocky substrate and reach 30 feet along the inside periphery of the kelp bed. At low tide, a plethora of plants and animals cling to exposed rocks or are trapped in submerged, shallow pools.

While the undersea life beckons everyone to discover its mysteries, the land above has a beauty all its own. The coastline along the Reserve is made up of pockmarked boulders, their edges pounded smooth by ages of wave action, tidal flow, and a sprinkling of winter storms. Growing at the base of these boulders, luxuriously long strands of sea grass appear as hula skirts of verdant velvet when submerged with the tides. Eons of erosion have masterfully carved a maze of rock formations to create an adventurer's dream of caves and swim-throughs.

This book will increase your enjoyment of the Ecological Reserve and its environs. Your window on the world below may be easily explored by donning a mask or a pair of goggles. The Reserve and its surroundings are conveniently mapped out for you, while the photographs and accompanying text describe and explain special behaviors and characteristics of the plants and animals.

Please become a partner in helping to protect marine life in the Reserve by not littering it and by not disturbing its plants and animals. Help preserve this priceless jewel and, in turn, derive maximum enjoyment from your visit.

The Ecological Reserve

The San Diego City Council created the Underwater Park in 1970 amidst concern that marine life was being damaged and in some cases depleted by uncontrolled diving. The Council dedicated 6,000 acres of submerged lands as an underwater park, from San Diego's northern border at Torrey Pines State Park to La Jolla Cove. While invertebrates like lobster and abalone would be protected, recreational fishing would be permitted. Responsibility for park maintenance was to be shared by the City of San Diego's Department of Parks and Recreation and the California Department of Fish and Game.

The following year, the Council created the Ecological Reserve within the Park as a "look but don't touch" area to be used as a public aquarium. Encompassing 514 acres (just over a mile and a half of coastline), the Reserve would range from the University of California's southern border at La Jolla Shores, follow the coast south and then bend west, until it reached Goldfish Point. With this Reserve, protection would not only cover all marine life, but also every inanimate object as well. In this way, submerged ancient Indian artifacts would remain undisturbed.

Ten years later, in 1981, the Council again extended the boundaries of the "look but don't touch" Reserve to include La Jolla Cove and its surrounding coastline. This additional half mile, bound by Alligator Head to the north and Goldfish Point to the east, is the area emphasized in this book.

Formation of the Reserve has been critical in reestablishing depleted marine life species, safeguarding the area's fragile ecology, and preserving the natural beauty of the shoreline. While park officials have been mostly successful in providing protection for the flora and fauna, community education must continue in order to prevent ongoing degradation from poaching and littering.

Rules and Regulations Governing the Ecological Reserve
(Enacted March 8, 1971)

Pursuant to Section 1580 through 1584 of Fish and Game Code, the San Diego-La Jolla Underwater Park Ecological Reserve is to be preserved in a natural condition for the benefit of the general public to observe native flora and fauna and subject to the following regulations:

"No person shall disturb or take any plant, bird, mammal, fish, mollusc, crustacean, reptile or any other form of plant life, marine life, geological formation, or archaeological artifacts..."

Enforced by the State of California Department of Fish and Game

History of La Jolla Cove

Knowledge of human activity around La Jolla Cove and its surrounding cliffs dates back over 10,000 years to the Native American tribes living along the shoreline. Their existence is well documented by the thousands of mortar bowls, anchor stones, fishnet weights, and charm stones found either washed up on the beach or discovered by divers.

Below the cliffs, in the vicinity of the Cove, are a series of seven caves lined up in a row. The Cove and cave areas were used as a summer fishing camp by a tribe of Indians who lived a distance away in the mountains. They fished all summer, dried their catch, and returned to the mountains for winter.

This history has fueled debate about the meaning of La Jolla's very name. Many people believe it is a Spanish translation of "The Jewel," so-named for the beauty of the area. But arguments persisting over 100 years insist that for this to be true, the Spanish spelling should be "La Joya." Local Indians, as well as the visiting mountain tribe described above, gave the name " 'mut la Hoya, la Hoya" to this strip of coast. The translation of "Hoya" is cave. The word was said in repetition to denote multiplicity, hence, "place of many caves." What is the truth? The debate continues, yet either explanation describes this natural treasure.

The little green lab

By the late 1800s, La Jolla Cove and its cliffside surroundings were dotted with cottages owned by the well-to-do or maintained as rentals and used by vacationers. Of course, some people came here to live permanently, and one of these colorful characters made his mark in local history.

Professor Gustav Schultz, a civil engineer from Germany, arrived in La Jolla in 1902. He loved swimming in the ocean, particularly by the caves. Wanting to find a way for more people to experience their beauty, he hired two workers to dig a passageway from the top of the cliff down to the westernmost cave, now known as Sunny Jim. After two years, the arduous job was completed. Eventually, other workers widened the opening and installed 133 stairs. The entrance to this corridor is found on Coast Boulevard inside the Shell Shop. To this day, for a nominal fee visitors can descend the stairs to experience this bit of history—and remain dry! Those willing to take the plunge can partake of a free view by simply swimming up to the cave's opening (or any of the other six caves).

A world-renowned institution also originated at the Cove. In 1907, on the land above the Cove, the Scripps family financed the construction of a tiny building known as Marine Biological Associates or "the little green lab." A small group of scientists collected and studied marine life here until 1910, when they moved to the beach at La Jolla Shores and eventually formed Scripps Institution of Oceanography. While the lab no longer stands at the Cove, near its place is the Shuffleboard Club, a small building rented by locals for meetings and social occasions.

The Riviera-like coast and beaches have meant many things to many people, from the Native Americans to present-day visitors. While the invasion of people has changed the land, La Jolla Cove and its surrounding terrain have survived as a reminder of the beauty and simplicity of life long ago.

Points of Interest

The Cove Located on the ocean, La Jolla Cove is less than a half-acre parcel of beach-park protected from the waves by protruding rock formations.

Takeoff Rock A conglomeration of boulders stretching from inside the Cove and arcing out around its mouth provides wonderful tidepooling opportunities at low tide. The outermost boulder is a barely submerged, flat-top rock called Takeoff Rock. You can often see people poised atop it as though they are magically standing on water.

Alligator Head This tip of land marks the northwest boundary of the Reserve. To reach Alligator Head, swim seaward from inside the Cove and round the bend to the left. Facing west from a distance, the slabs of rock form a profile likeness of an alligator in repose staring out to sea.

Emerald Cove Swimming seaward out from the Cove, bear right (you will be heading east) to reach this rocky cove. Swim almost up to its shore and look toward the knoll on the left. There you will see a large opening called the Swim-Through.

Swim-Through At first glance this structure looks like a cave, but it is actually a swim-through. Enter only when water is calm to avoid turbulent conditions within. Emerge carefully out the other side, which is directly under Goldfish Point.

Goldfish Point This small knoll was named for the numbers of garibaldi or "ocean goldfish" observed in the water below. The outer end of this knoll is frequented by jumping and diving enthusiasts who make the 30-foot leap into shallow waters directly outside of the Swim-Through. (Goldfish Point can also be reached by land. Just hike down from the Shell Shop on Coast Boulevard.) Continue your swim around the knoll to the right to reach the caves.

The Caves Seven tidal caverns, believed to have been created by wave action and mineral reaction over time, convene at the base of rugged 300-foot cliffs. Sunny Jim, the first cave you reach, was named for a popular 1890s comic strip character who had a Kewpie-style topknot of hair. Looking out from inside the cave, you can see a silhouette resembling Sunny Jim. Sunny Jim is also the cave to which steps descend from within the Shell Shop above. Each cave has its own personality, so explore them all when the water is calm. Just beyond the seventh cave you will come to Devil's Slide.

Devil's Slide Beginning at the small, crescent-shaped, rocky beach just past the easternmost cave, Devil's Slide bends northward following the stretch of coastline. The name describes the large expanse of crumbling cliffs hovering overhead. Continue hugging the coast to remain in shallow, reefy water that winds its way to La Jolla Shores, where it ultimately gives way to sandy beach.

How to Maximize Your Sightings

While some of what you see in the Ecological Reserve will depend upon luck, there are a number of things you can do to maximize sightings of marine life.

Take a different approach when looking for a plant or animal. You may not realize that you need to do more then just "look around" for the marine life that you hope to see. Pay close attention to each description of "Where Found" and "Feeds On." Look carefully for those locations and food sources. Now you won't be *hoping* for a sighting but actively and systematically tracking down the subject. For example, you will not find free-swimming grunion under rocks or bottom-dwelling lobsters near the water's surface. Grunion are found schooling just below the surface, while lobsters are sequestered in dens under rocks.

Keep in mind that most animals survive by camouflage (unlike the neon-orange garibaldi) so do not expect them to visually jump out at you. Instead, focus on one small area and be patient. Your perseverance will surprise and reward you when something suddenly moves that you thought was just a rock or a ragged piece of seaweed.

By using these simple techniques, you will dramatically increase your chances of seeing marine life in its native habitat.

Picture Key to Fins of a Fish

To facilitate fish identification, it is helpful to know fin terminology.

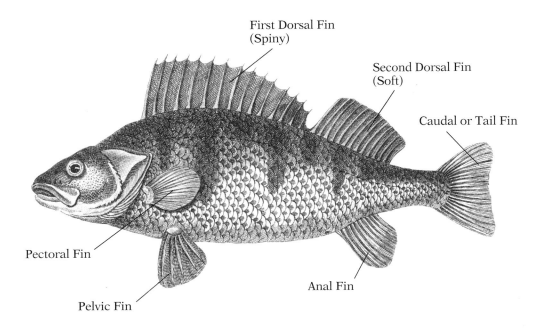

First Dorsal Fin
(Spiny)

Second Dorsal Fin
(Soft)

Caudal or Tail Fin

Pectoral Fin

Pelvic Fin

Anal Fin

Growing end; 3 inches long

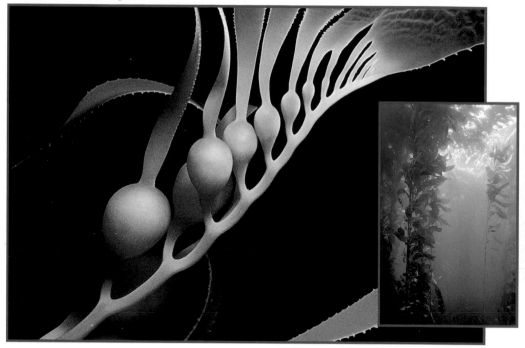

Appearance: From street level above the Cove, brown plants are seen as a shiny, slick, snarled mass breaking the surface of the water.

Size: Length from seafloor to surface—15 to 90 feet.

Where Found: The kelp bed is 200 yards north of the Cove. Some isolated plants are scattered around Emerald Cove and the caves.

Feeds On: Sunlight and organic matter (photosynthesizes).

Giant Kelp
Macrocystis pyrifera

Capable of growing up to 2 feet per day, the giant kelp is the fastest growing plant in the world. Forming the equivalent of forests found on land, a kelp bed shelters a tremendous density and diversity of marine life. How does this "underwater redwood tree" support its own ponderous structure? The secret lies within its oval bulbs, filled with gases (including carbon monoxide) to permit flotation. Look for growing ends within the tangled fronds. They demonstrate kelp reproduction. A single blade (leaf) splits, forming individual stipes (stems), air bulbs, and blades. Each newly formed kelp continues growing until its blade is ready to split and begin the reproductive cycle once again. Norris's top snail (p.24) may be found everywhere along the plant.

Feather Boa Kelp

Egregia menziesii

Because its air bulbs are a quarter of the size of the giant kelp's (p.12), the feather boa has less flotation; thus it may be found stretched out or in a knotted, tangled heap on the ocean floor. While an individual strand is capable of a vertical stance, it still manages to look more like a limp strand of spaghetti. Historically, this seaweed has been used as garden fertilizer. Look for the minute-sized snail, called a kelp limpet (p.22), that lives and feeds solely on the boa.

Appearance: A chocolate-brown plant with two radically different appearances. The healthy growing phase has small, laterally protruding air bulbs and flattened blades (leaves) placed opposite its stipe (stem). Resembling a feather boa scarf, an older plant has shriveled air bulbs and its blades degrade into fuzzy masses.

Size: From 10 to 20 feet in length.

Where Found: Attached to rocks and boulders, along reefy and sandy areas.

Feeds On: Sunlight and organic matter (photosynthesizes).

Braided Hair

Plocamium coccineum var. *pacificum*

Appearance: A deep-red and branching alga. Delicate tufts of featherlike growth zigzag off a supple, slender axis.

Size: To 12 inches in height.

Where Found: In tide pools and on rocky surfaces below the surface.

Feeds On: Sunlight and organic matter (photosynthesizes).

Braided hair is considered the most beautiful local seaweed for its color and frilly form. The alga's red pigment masks its underlying green tint from chlorophyll. Its flexibility provides excellent protection against wave shock and strong surge. When the alga dies, the sun bleaches it pale yellow, leaving clumps of faded braided hair clinging to rocks surrounded by its thriving brethren. The giant kelpfish (p.54) resides amidst the alga's fanlike configurations and is patterned in the same alluring hue. Inspect closely to discern the fish from the seaweed.

Surf Grass

Phyllospadix scouleri

Not a primitive plant like kelp or other algae, surf grass is a true rooted, flowering plant. Often mistaken for "eel grass" (a similar plant found in calm bays and lagoons), surf grass differs in its tremendous anchoring power, a critical asset in warding off the harsh conditions of surf and surge. Thickly draped over rocks when exposed at low tides, this grass becomes alive when covered by water at higher tides. Under the surface, surges of water whip and whirl the green blades. The giant kelpfish (p.54) may be seen within the swirling grass, sporting the same vibrant grass-green color. Baby spiny lobsters (p.29), less than an inch long, also can be found clinging to the grass blades.

Appearance: Narrow, compressed leaves appear as brilliant, emerald-green blades growing individually in dense, lush mats.

Size: Single strands to 4 feet in length.

Where Found: Attached to rocks and boulders just below the water's surface in areas of surf or active water motion. Some grass is exposed to air at low tides.

Feeds On: Sunlight and organic matter (photosynthesizes).

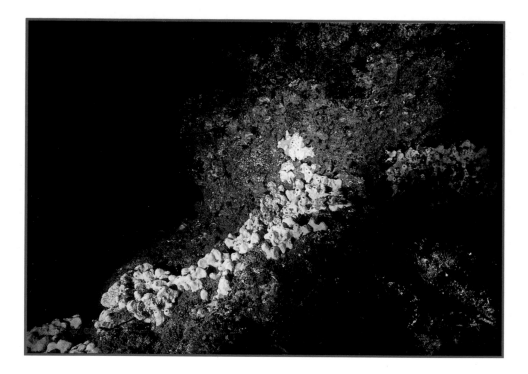

Appearance: Flashy yellow color and craterlike projections make this an easily identifiable sponge.

Size: Forms an irregular crust about 1 1/2 inches in height by 15 inches in diameter.

Where Found: Mostly under ledges where it prefers darker accommodations.

Feeds On: Bacteria and dead particles.

Sulfur Sponge
Aplysina fistularis

Melanin, the dark pigment found in skin and hair, is found in the tissues of the sulfur sponge as well. The sponge also harbors poisonous substances within its tissues to deter potential predators. Sponges are actually primitive colonial animals with many tiny individuals living together as one. Tiny holes across the sponge draw water down canals where food and oxygen can be absorbed. Snorkel down to an area where the rocks are undercut and look beneath the ledges. You are certain to see one of these vivid, encrusting colonies. An eagle eye might also spot the tiny sea snail, Tylodina (p.27), which generates its matching lemon color by feeding exclusively on this sponge.

Aggregating Anemone
Anthopleura elegantissima

Its green color is found only on those anemones exposed to the sun and is due to the presence of algae growing within its tissues. A specimen spending its life away from sunlight is white. When prey touches the anemone tentacles, venom is released, paralysis occurs, and the anemone brings the new food into its mouth. During low tide, the exposed anemone folds its tentacles in for safety. Its stalk, now revealed, is covered with sucking discs to which bits of broken shells and gravel stick for protection. This acquired armor makes the anemone less vulnerable to drying out and to attacks from hungry birds.

Appearance: Deep-green or bluish in color with fine purple stripes radiating out from the mouth to the tentacles. Tentacle tips are purple. Its flowerlike "face" is supported by a stalk that attaches to rocks.

Size: To 4 inches in diameter.

Where Found: Mostly in tide pools around the exposed boulder areas inside the Cove. It is found singly or in congregations where it burrows into crevices or anchors to rock faces and rocks embedded in sand.

Feeds On: Fishes and crustaceans.

Appearance: Worms have a crown of lavender tentacles and live in self-created tubes of cemented sand grains. Tube masses resemble honeycombs and harbor a single worm per tube. When uncovered at low tide, worms contract into their tubes.

Size: Worms to 2 inches in length; "sandcastles" to several feet across.

Where Found: On and under rocks exposed to air at low tide in tide pool area and around shallow reef areas where rocks meet sand.

Feeds On: Microscopic animals that become trapped amongst the tentacles.

Sand-Castle Worm
Phragmatopoma californica

This industrious worm traps sand grains from the surging water to create its protective tube. A special pocketlike organ behind the mouth accepts the particles and stimulates secretion of mucus, a liquid-cement mixture it uses to build the wall of the tube. These unmistakable sandcastles are very delicate and can be easily crushed. Find one under water and look closely for the pale purple tentacles protruding from the opening of each "honeycomb." The knobby sea star (p.31) is a predator.

Mossy Chiton

Mopalia muscosa

To create a depression, the chiton grinds away at the rock using its radula (a tongue dotted with many tiny teeth). Bending into the depression is no problem for this unusual mollusc because its eight-jointed shell permits great flexibility. Dense bristles form a tight suction to the rock, permitting exposure to air for periods of time while preventing desiccation. A nocturnal feeder and homebody, the chiton departs its depression during darkness to graze but always returns home by daylight. Chitons use the same depressions generation after generation, so many of these pits may be a thousand years old or more!

Appearance: Oval-shaped shell has greenish-brown plates usually encrusted with plants and/or animals. Thick, yellow-brown bristles surround its periphery.

Size: To 2 inches in length.

Where Found: In depressions on boulders at the low tide line, where water submerges the rocks and then exposes them to air.

Feeds On: Red and green algae, and scavenges on animal matter.

Appearance: Reddish-brown, oval, fairly flat shell usually concealed by heavy algal growth and encrusting animals. Distinguished by 5 to 7 open holes on one side of the shell and numerous grey to light-green sensory tentacles jutting out along the bottom of the shell. Ruffly mantle extending out below the shell is mottled green and brown.

Size: To 10 inches in length.

Where Found: Attached to rocks under ledges and between crevices.

Feeds On: Algae, especially the giant kelp (p.12).

Green Abalone
Haliotis fulgens

The abalone firmly attaches itself to rocks by suction grooves on its muscular foot. Small holes along one edge of the shell are required for respiration and excretion of waste or reproductive products. A hemophiliac, with no clotting mechanism, the abalone will bleed to death even if only slightly injured. Look for "rocks" with suspicious moving tentacles to find this elusive mollusc. Since the abalone is California's most delicious marine snail, heavy poaching in the Reserve has prevented any increase in its already minuscule numbers. See giant keyhole limpet (p.21), a related species.

Giant Keyhole Limpet

Megathura crenulata

Unlike other snails, the giant keyhole limpet's body is too large to hide under its shell. The body must extend around the shell on all sides. Only a "keyhole" opening is exposed at the top of the shell so that water can exit after passing over the gills. Coastal Indians used the shells as wampum (money). Classified in the same group as the abalone, the giant keyhole is eaten in Japan, but its strong flavor and rubbery texture make it unpalatable for the average American's taste.

Appearance: Cream, mottled-brown, or black body extends over a volcano-shaped shell, obscuring it almost completely.

Size: Shell to 5 inches in length; body to 10 inches in length.

Where Found: On exposed rocky surfaces.

Feeds On: Algae.

Appearance: Reddish-brown, flattened shell looks polished and shiny.

Size: To 3/4 inch in length.

Where Found: On the stipes (stems) of feather boa kelp.

Feeds On: Mainly the feather boa kelp (p.13) stipes on which it lives.

Kelp Limpet
Acmaea insessa

Settling onto the thick stipe of the feather boa kelp, the kelp limpet munches away, eventually creating a depression in the kelp. When the limpet moves to another location, the distinct imprint is left behind. Inspect any stipe and you will easily find empty "footprints" of long-gone limpets along with living limpets secure in their custom-made impressions.

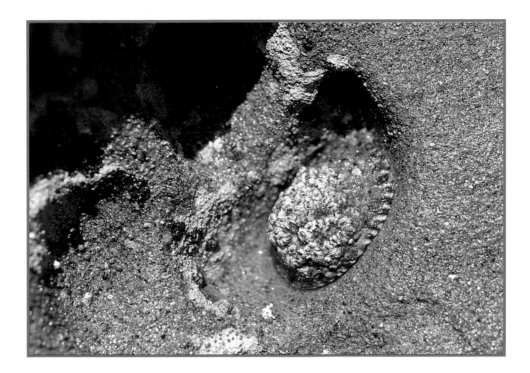

Owl Limpet

Lottia gigantea

Often mistaken for a baby abalone, this limpet is notable for its interesting territorial behavior. Each owl limpet grazes on a 13-inch-square film of algae growing near its home. Other animals are not allowed to settle there. This beast rids itself of potential interlopers by rasping them off with its radula, a sandpaperlike tongue, or by pushing them, using its shell as a bulldozer. In this way, a large patch of algae grows dense enough for the limpet's continuous consumption without competition from other grazers.

Appearance: Brown shell has radiating white bands. Oval and low in profile, the shell is often adorned with tiny hitchhikers such as the barnacles in the above picture.

Size: To 3 1/2 inches in length.

Where Found: On rocks and boulders nestled into a self-inscribed depression on the rock. Depression is created by the owl limpet's gritty tongue persistently grinding away on the rock. Frequently exposed to air, it is easily found while tidepooling.

Feeds On: Algae.

Appearance: Smooth, thick, light-brown shell, sometimes burdened with algae and barnacles. Easily identified by its striking orange-to-red protruding foot.

Size: To 2 inches in diameter.

Where Found: On brown seaweed, especially the giant kelp.

Feeds On: Seaweed.

Norris's Top Snail
Norrisia norrisi

A unique and systematic lifetime agenda exists for this top snail. Beginning at the base of the plant, the snail crawls vertically up the stipe, grazing as it goes. When it reaches the top of the plant, it literally falls off, hits the ground, orients itself, and heads back to the same plant to repeat the process. Look for it throughout the dense foliage of the giant kelp (p.12).

Wavy Top Turban

Astraea undosa

When threatened, the turban retreats into its shell, slamming its "door" as it flees. This pearly white protective door, called an operculum, is teardrop-shaped and lightly engraved with two wavy ribs. Polished and free of algal growth, the calcium carbonate concretion sticks tightly to the back of the snail's foot. Often washed up on the beach, turban doors are frequently mistaken for shell fragments. The wavy top turban is a good food source for the two-spotted octopus (p.28).

Appearance: Light-brown, large pyramid-shaped shell is engraved with vertical wavy ridges. Shell is camouflaged with encrusting marine growth, usually red algae. Extended foot is white and brown in a mosaic pattern. A slim, white eye stalk projects from the bottom of the shell just above the foot.

Size: To 4$^{1}/_{2}$ inches in height and 4 inches in diameter at the base of the shell.

Where Found: In rocky areas on the ocean bottom amidst red algae-covered boulders. Many may be found around the cave areas.

Feeds On: Seaweed.

Appearance: Body is orange. Heavy shell is off-white to pale green and often overgrown with algae or other organisms. Shell has a high spire of 7 to 8 whorls and a long curving canal with 8 to 9 nodes ornamenting each whorl.

Size: Shell to 6 1/2 inches in height.

Where Found: Along the rocky areas and on the floor of the kelp bed. Empty shells may be washed into the tide pool area.

Feeds On: Dead or injured animals it finds on the seafloor.

Kellet's Whelk
Kelletia kelletii

One of the largest and longest-living snails found in Southern California, a 3-inch sized individual may be 7 or 8 years old. It often invades lobster traps and may kill the captive crustaceans. Frequently seen feeding with the predatory knobby sea star (p.31), it shares common food items like abalone and clams. In late spring, large breeding groups congregate on the rocky floor of the kelp bed to lay watermelon seed sized eggs that resemble cream-colored buttons. Whelk shells have been found in local Indian middens (refuse heaps) and are thought to have been used for trumpets or bird calls.

Sulfur Sponge Tylodina

Tylodina fungina

After feeding on its host sponge, a Tylodina absorbs the sponge's toxic components and flamboyant color. The defensive chemicals are stored in special glands, rendering the snail as distasteful as the sponge. The lurid color distributed throughout the snail's body wards potential predators away from its repulsive taste and toxic effects. Hermaphroditic, each Tylodina individual contains both male and female sexual organs and can mate with any other Tylodina.

Appearance: Brilliant yellow color with horns protruding atop head. External cone-shaped shell presents a limpetlike appearance.

Size: Almost 1 inch in length but only 1/4 to 1/2 inch in height.

Where Found: On host sulfur sponge (p.16) and red algae.

Feeds On: Sulfur sponge.

Appearance: Changes color to match its background. Below the true eyes, two dark, large ovoid spots display a deep-blue color.

Size: To 3½ feet, including body and outstretched arms.

Where Found: Between rocks and in holes. (A pile of crab shells outside the opening of a hole often designates an octopus den.) Occasionally out swimming, or perched atop a rock.

Feeds On: Crustaceans, molluscs such as the wavy top turban (p.25), and small fishes.

Two-Spotted Octopus
Octopus bimaculatus

A highly developed camouflage animal and quick-change artist, the octopus is capable of varying color and texture at speeds surpassing even the chameleon. This nocturnal mollusc usually waits until dark to emerge and forage for food. Highly intelligent, it is extremely sophisticated in capturing prey. After enveloping its victim in a loving embrace, it immediately releases toxic saliva into the water or directly into the victim via its sharp beak. The captive is instantly paralyzed and ready to eat. Predators include the moray eel (p.35) and cabezon (p.39).

California Spiny Lobster

Panulirus interruptus

Using its extra-long antennae, this scavenger deploys a sweeping motion to chase off other animals attracted to its prey or to frighten enemies away. If this defensive mechanism fails, a structure at the base of each antenna broadcasts an alarming grating noise. To find one of these nocturnal dwellers, look under ledges or in holes with protruding antennae. Should you encounter a lobster wandering in the open, this shy creature will curl its tail under and flap away in an instant. It is nicknamed "bug" by divers because its segmented shell, multiple appendages, and long antennae resemble a cockroach of gigantic proportions. The largest lobster documented weighed a hefty 28 pounds.

Appearance: Red to orange coloration. Spiny projections stud the carapace (upper shell) and the sides of its tail. The two antennae are twice as long as its body.

Size: To 2 feet in length across carapace.

Where Found: Under ledges and in holes, often in groups of 3 to 6. May also sit out in the open atop rocks and surf grass (p.15).

Feeds On: Plants, animals, and decaying matter.

Striped Shore Crab
Pachygrapsus crassipes

Appearance: Dark stripes are spread over a square-shaped body colored purple, green, or red. Has eight dark legs, and a large set of claws tinted pinkish on top and white at the tips.

Size: To 2 inches across carapace (upper shell).

Where Found: On the rocks exposed to air and just below the waterline.

Feeds On: Algae and microscopic plants growing on rocks but may also consume snails, other crabs, and limpets.

Semiterrestrial, hardy, and tough, this crab can survive out of water for almost three days (if shaded) by retaining a drop of water in its gill chamber and reducing its breathing rate. This clever feat, along with its ability to withstand fresh or salty water and fluctuating water temperatures, makes it the most versatile and common crab known on the West Coast. When hassled by a human, it thrusts out its claws in preparation for battle and will happily give you a painful nip. Predators include the sea gull, anemones such as the aggregating anemone (p.17), and various fishes.

Knobby Sea Star

Pisaster giganteus

While the backs of many other animals become riddled with plant and animal growth, why does the sea star remain conspicuously clean? Because its entire surface is covered with tiny pinching organs that grasp and crush, anything landing on its back is immediately destroyed. This voracious predator also strikes fear into abalone, limpets, and other snails. It often feeds with the scavenger, Kellet's whelk (p.26). To find this animal, swim out to the kelp bed. When the surface-to-seafloor visibility is good, you will easily see many examples of this sea star 25 feet below, scattered along the ocean floor.

Appearance: Nubby white spines set in an irregular pattern cover the top of its five-armed form. A circle of blue illuminates the base of each projecting spine.

Size: To 22 inches in diameter.

Where Found: On the rocky seafloor of the kelp bed.

Feeds On: Barnacles and molluscs such as clams, abalone, snails, and chitons.

Leopard Shark

Triakis semifasciata

Appearance: The top and sides of its gray body are adorned with thick black, elongated spots. Underbelly is white. Moves in a snakelike motion.

Size: To 7 feet in length.

Where Found: In sand off the beach at La Jolla Shores and occasionally cruising the Cove's rocky areas or kelp bed.

Feeds On: Clams, octopuses, crabs, spiny lobsters (p.29), bony fishes, and bat rays (p.34).

The leopard shark is not dangerous and is, in fact, fearful of divers' bubbles or swimmers' movements. During the summer, many congregate on the sandy bottom at La Jolla Shores to breed, but an occasional one streaks by the Cove. After a 12-month gestation period, females give birth to litters of 4 to 29 offspring. Juveniles grow up to 4 inches a year and have an average life span of about 25 years. The leopard shark ranges from Oregon to the central coast of Mexico, where it dwells in shallow waters to depths as great as 300 feet. In Southern California, fossil leopard sharks have been found in deposits over a million years old.

Round Stingray

Urolophus halleri

The most common of California stingrays, this perceived demon sends shivers down bathers' spines. You should be safe from the sting of this shy creature because the steepness of the beach entry at the Cove and its abundance of rocks are not conducive to the ray's favorite habitat. It prefers a large expanse of sand, as found off the beach at La Jolla Shores, where it buries itself up to its eyeballs and hopes for peace and quiet. (Honest!) Stingray locations around the Cove include the deeper sandy flats like those between the caves and the orange-striped swim buoy. Females give birth to as many as six baby stingrays.

Appearance: Mottled-brown, gold, or gray coloration on its small, circular body. Short, thick tail harbors a long spine that reaches toward the tip of the tail.

Size: To 22 inches in length.

Where Found: Camouflaged on the sandy bottom and undulating over the rocky areas.

Feeds On: Bottom animals such as worms, shrimps, and crabs.

Bat Ray

Myliobatis californica

Appearance: Blackish or brown on top with white undersides. Marked by a protruding head and distinct face. Base of its long, whiplike tail harbors a short stinger, making it a less effective striking organ than that of the round stingray's (p.33).

Size: Female to 6-foot wingspan; male to 2-foot wingspan.

Where Found: Usually on sandy bottom off La Jolla Shores' beach but frequently sighted at the Cove flying over the rocky bottom, or winging its way through the kelp forest.

Feeds On: Oysters, clams, crabs, shrimps, and fishes.

This stingray is infamous for its impressive grinding teeth. An adult is capable of exerting a walloping 150 pounds of pressure through its mouth, which makes crushing a clam or oyster shell as easy as chewing Jell-O. Many people confuse the bat ray with a manta ray. Aside from other physical differences, a manta's wingspan sprawls 18-feet wide and its weight swells to 2,300 pounds. This constitutes an average 5-fold greater wingspan, and a 2,100 pound difference in weight compared to the relatively dainty 200-pound bat ray.

California Moray Eel

Gymnothorax mordax

Although the moray's mouth lies agape in a menacing manner, this expression is merely a breathing mechanism that allows water to be washed across its gills. Nevertheless, its species name, *mordax,* means "prone to bite." When threatened or harassed, this eel may cause injury. Keep your hands out of any opening in the rocks—it may be the home of a moray. Living to thirty years or more, the moray may be found residing benevolently in the same hole as the spiny lobster (p.29). An aggressive predator, the eel descends into an octopus' den to root it from its hiding place much like a gopher snake wiggles into a gopher burrow.

Appearance: Brown, smooth skin. A fine milky-blue color outlines its eyes. Snakelike body. Mouth large with easily visible, pointed white teeth.

Size: To 5 feet in length.

Where Found: Deep within clusters of rocks, where only its head protrudes.

Feeds On: Crustaceans, fishes, sea urchins, and octopuses.

Pacific Sardine

Sardinops sagax

Appearance: Elongated blue or green-backed fish with a silvery, reflective belly and a longitudinal row of 7 or 8 black dots on each side.

Size: To 16 inches in length; in the Reserve, only those a few inches long are found.

Where Found: Just below the surface in large schools.

Feeds On: Microscopic plants.

The sardine's vast range is truly titanic, extending from Kamchatka, Russia, to Southeast Alaska and south to Guaymas, in the Gulf of California. A sardine fishery begun on this coast during World War I continued to escalate in size until, at its peak, up to 800,000 tons of sardines were landed annually. By the late 1940s, catches were drastically reduced due to overfishing, and the industry died. Since then, sardine numbers have increased considerably. Predators include a wide range of fish, birds, marine mammals, and humans who use them for bait.

Topsmelt

Antherinops affinis

If you see a fish leaping out of the water, it is most likely a topsmelt. To escape larger predatory fish such as mackerel, the topsmelt takes flight to save its life. Spawning occurs in spring and summer when the female, closely followed by several males, deposits sticky egg strands onto various underwater plants. Once laid, the eggs are immediately fertilized by the males. Highly adaptable, the topsmelt thrives in conditions from fresh water to salinity three times saltier than seawater!

Appearance: Elongated body is bright green above and silvery below with a midline greenish stripe. Its anal fin begins directly (or almost directly) below the first dorsal fin.

Size: To 14½ inches in length.

Where Found: Seen in handfuls or in loose schools just below the water's surface. Prevalent inside the Cove, around the kelp bed, and along sandy areas.

Feeds On: Microscopic animals, algae, and insect larvae.

Black & Yellow Rockfish

Sebastes chrysomelas

Appearance: Blackish or brown-ish body that is smooth and appears scaleless. Decorated with yellow blotches and spotting on its back.

Size: To 15½ inches in length.

Where Found: In rocky areas inside the mouth of the Cove, snuggled under ledges and inside crevices. Also found in the kelp bed.

Feeds On: Crabs, shrimps, and occasionally fishes and octopuses.

While the rockfish is not quite a chameleon, it can vary its color and pattern slightly to better blend into its surroundings. It is most prevalent in shallow water but can be found to depths of 120 feet. Because the dorsal, anal, and ventral spines are mildly venomous, a puncture wound can cause minor injury. The rockfish is ovoviparous, meaning that a female is fertilized internally by a male. Once fertilized, the mother incubates the million-plus eggs until the embryos are ejected—in the form of gelatinous, free-floating balloons. Experts are mystified over the multitudinous variations of *Sebastes*; there are an extraordinary sixty-nine species in all.

Cabezon

Scorpaenichtlys marmoratus

The cabezon's substantial and bulbous head easily distinguishes it from any other fish. Peculiarly placed eyes permit a panoramic view, valuable for spotting prey and predators. Staunchly monogamous, a cabezon couple returns to the same nesting place every year. The female lays up to 100,000 eggs, and the male guards them until hatching. A large cabezon can swallow an abalone whole, digest the meat, then spit out the shell.

Appearance: Mottled-red males or mottled-green females, but either can change color to mimic background. Body is scaleless. Dorsal fin is prominent when raised. Head is large and lips are full and fleshy. Branched, hornlike protrusions perch above its eyes, which are placed wide on the head.

Size: To 3 feet in length.

Where Found: A bottom-dwelling fish that hides among the rocks or sits atop algae-covered boulders.

Feeds On: Shrimps, crabs, abalone, and small fishes.

Kelp or Calico Bass
Paralabrax clathratus

Appearance: Mottled in brown to olive-yellow above and yellow to gray-white beneath. Sensational emerald-green eyes. Older males have yellow lips.

Size: To 26 inches in length.

Where Found: Abundant around giant kelp (p.12), surf grass (p.15), and rocky areas. Seen singly or in loose groups.

Feeds On: Small fishes, crabs, and some sea stars.

This dour-looking fish simply cannot be rushed. It moves at the speed of a gently swaying kelp frond. The bass' calm nature makes for a relatively approachable fish, since it is not easily spooked. As a matter of fact, when confronted by a human, the bass will most likely just "hang" motionless and stare—glumly. Kelp bass may live up to thirty years. An extremely popular sport fish, it is taken mainly by anglers on California charter boats. Skeletal remains have been found in Indian middens (refuse heaps).

Sargo

Anisotremus davidsoni

When seen in schools of fifty or more, this fish may be one of the most beautiful sights around the reef. Very young juveniles school within the Reserve beginning in late summer to early fall. The sargo is commonly taken by anglers, usually as an incidental catch. When removed from the water, it makes a piglike grunting sound. The sargo is also found in the Salton Sea (east of San Diego), where it was introduced in 1951.

Appearance: Silvery-gold adult wears a prominent, thick black vertical stripe draped over the upper sides of its oval, compressed body. The contrasting juvenile has several dark, horizontal stripes and an elongated yellow body.

Size: To about 17 inches in length. Juvenile loses horizontal markings after growing to 4 inches.

Where Found: In midwater and close to the bottom near kelp or rocks. Seen singly, in pairs, or schools.

Feeds On: Tiny invertebrates such as isopods and shrimps.

Appearance: Dark- or olive-green back and green sides. It has 1 or 2 conspicuous white spots centrally located on each side of its back by the dorsal fin. Eyes are a beautiful opalescent blue.

Size: To 26 inches in length.

Where Found: Around rocks and over large boulders. Found singly, in pairs, or in small schools of 15 to 20.

Feeds On: Algae and small crustaceans living upon the algae.

Opaleye
Girella nigricans

Ubiquitous throughout the Reserve, the opaleye is often found in loose schools, head down, plucking at short tufts of algae on boulders. Adults spawn from April to June, and the young opaleye enter the tide pool area from June to early winter to grow in the relative safety of the shallow pools. At low tide, you can find this hyperactive fish darting around, the white specks atop its back allowing for easy identification.

Zebraperch

Hermosilla azurea

Despite its common name, the zebraperch is not a perch at all. It is a member of the sea chubs, which include the opaleye (p.42) and halfmoon (p.44). Between its peculiar turned-in mouth and referee attire, the zebraperch is one of the more easily recognizable fish in the park. A school of adults may be seen hovering a few feet away from a school of juveniles, but the schools never interact. The zebraperch is often found inside the Cove in just a few feet of water.

Appearance: Yellowish cast with 11 or 12 dark, vertical bars stretching over its body, punctuated by a bright patch of blue lying just behind its gills. Turned-in mouth somewhat resembles a parrot's beak.

Size: To 17 inches in length.

Where Found: Around kelp, sand, and rocky bottoms, swimming singly or in small schools.

Feeds On: Algae and small invertebrates.

Halfmoon
Medialuna californiensis

Appearance: Grey-blue with a distinctive halfmoon-shaped tail and extended protruding dorsal and anal fins.

Size: To 19 inches in length.

Where Found: Around reefs and in the kelp bed.

Feeds On: Seaweed and small invertebrates.

Prevalent all along the coast, this beautiful fish may be found singly or in schools, particularly in the kelp bed. Predators include the kelp bass (p.40), sea lion (p.58), and harbor seal (p.59). There is a small sport fishery for halfmoons, mostly for recreational anglers. The halfmoon ranges from Vancouver Island, Canada, to the Gulf of California.

Kelp Surfperch

Brachyistius frenatus

One of the nineteen species of saltwater perch that inhabit California waters, the small kelp surfperch ranges from Vancouver Island, Canada, to Bahia Tortugas in central Baja California. Some kelp surfperch are "grooming" fish; they enjoy a good meal from the dead skin and parasites picked off another fish's body, particularly the black-smith (p.50).

Appearance: Brassy-golden with a small, upturned, pointed snout.

Size: To 8^1/$_2$ inches in length, but 3-inch juveniles abound.

Where Found: Under and around kelp canopy as well as on kelps in rocky reef area. Found singly or in schools.

Feeds On: Eats microscopic animals, especially in kelp.

Appearance: Burnt-orange color. Several dark bars on sides and a turquoise bar at the base of the anal fin. Prominent yellow-orange lips.

Size: To 15$\frac{1}{2}$ inches in length.

Where Found: Reefs and shallow kelp beds.

Feeds On: Fishes, tiny shrimps, and crabs.

Black Surfperch
Embiotoca jacksoni

While almost all other marine fish scatter eggs outside the body, the remarkable perch spawn live baby perch. Even more amazing is the male of the species, which is sexually mature at birth. Feeding is unusual, too. After sucking in a clump of algae and rock, special throat muscles are used to separate out tiny shrimps, crabs, or fishes; the rest is simply ejected. The black surfperch may be found from shallow waters down to 150-foot depths.

Walleye Surfperch

Hyperprosopon argenteum

Small to large schools of this surfperch may be found tightly packed together circling the reef. The male engages in an aggressive swooping courtship display. The female's gestation period is 5 to 6 months, with births occurring in April. Fry (the young of fish) number 5 to 12 and are a strapping 1½ inches long. The walleye is considered important to both the commercial and sportfishing communities. Native Americans included walleyes as part of their diet. Fossils of this fish have been found in deposits over one million years old.

Appearance: Highly reflective, silvery color. Tips of pelvic fins and edge of tail are black. Has a huge pair of eyes.

Size: To 12 inches in length.

Where Found: Midwater over rocky areas and around kelp.

Feeds On: Tiny crustaceans, such as krill, and small fishes.

Appearance: Silvery-blue body. A thin black line rims the base of the dorsal fin. Fins are yellowish.

Size: To 12½ inches in length.

Where Found: In loose schools in very shallow water over sandy reefs and in the kelp bed.

Feeds On: Bottom organisms such as crabs, and worms.

White Surfperch
Phanerodon furcatus

The prolific white surfperch female bears as many as thirty-three young at a single time. This surfperch may be found from the surface to cavernous depths of more than 200 feet. Living to seven years, it is a commonly fished species, usually taken from piers and jetties. Aside from humans, predators include the harbor seal (p.59).

Male guarding egg mass

Juvenile; 1¹/₂ inches long

Garibaldi

Hypsypops rubicundus

This damselfish lives individually and is highly territorial as an adult. Threatening animals (including divers!) that approach its domain provoke the garibaldi to emit a clearly audible thumping noise as a warning to stay back. The juvenile's blue markings may exist to signal permission to enter into an adult's protective territory. Reproductive rituals include the male "building a nest" by grooming a lawn of short, red algae growing against the slanted surface of a rock. The female deposits bright-yellow eggs on the algae in an oval mass, but the male is sole protector of the eggs until hatching. Nesting occurs March through July. The garibaldi is protected by law and cannot be taken for sport or commercial purposes. See black-smith (p.50), a related species.

Appearance: Adult is bright-orange to orange-red. Juvenile is streaked and spotted luminous blue atop a bright-orange background.

Size: To 14 inches in length. Juvenile loses blue markings after growing to 6 inches.

Where Found: In the kelp bed, and along reefs near openings in rocks where it can retreat to safety.

Feeds On: Seaweed, sponges, bryozoans, worms, and crabs.

Juvenile; 2 inches long

Appearance: Adult is dark blue with black spots on the posterior half of its body. Juvenile is a rainbow of colors; the front half is gray-blue, the back half is yellow-orange and violet, the tail is golden.

Size: To 12 inches in length. Juvenile loses special markings after growing to 2 inches.

Where Found: In midwater above the rocks and reefs and especially around kelp.

Feeds On: Plankton (microscopic plants and animals).

Blacksmith
Chromis punctipinnis

Late summer brings sizable schools of juveniles to the Cove. Along with the garibaldi (p.49), the blacksmith is the only other damselfish found in these temperate waters, since damsels are usually associated with coral reefs. Both damsels can be seen sleeping under ledges at night. Blacksmith differ from garibaldi in that they school. Both fish are alike in behavior during spawning season, when the male leaves the school to groom a nest. The blacksmith female lays salmon-colored eggs. Senorita fish (p.51) may clean the bodies of quietly patient blacksmith in a behavior that benefits both. The senorita relieves the blacksmith of tiny parasites and consumes a tasty meal.

Senorita

Oxyjulis californica

The most prevalent "groomer" or "cleaner" fish off the Southern California coast, the senorita may be found individually or in small numbers. Groups of senoritas set up a grooming station (especially during the summer), and other fish drop by to be cleaned. Just before sunset and when threatened, the senorita may dive headfirst into the sand for shelter. At sunrise, or once the threat is over, it emerges. From late summer and throughout winter, juveniles are seen swimming about in large schools.

Appearance: Its upper half is yellow-orange and its lower half is cream-colored. Body is cigar-shaped. Tail fin is stamped with a large black spot. Close inspection reveals tiny teeth protruding out of its mouth.

Size: To 10 inches in length.

Where Found: Rocky areas and kelp bed. May be cleaning (picking) parasites from the bodies of sargo (p.41), opaleye (p.42), surfperches, and garibaldi (p.49).

Feeds On: Tiny crustaceans, worms, larval fishes, and parasites.

Female

Appearance: Body is cigar-shaped. Male is greenish with a vertical black bar over its head. Female is mostly pinkish to salmon colored; the top of its back is covered in black spots.

Size: To 15 inches in length.

Where Found: Nosing around on the rocky bottom, and grazing in sandy, gravelly areas.

Feeds On: Crabs, snails, and algae.

Rock Wrasse

Halichoeres semicinctus

The rock wrasse is a "sequential hermaphrodite," meaning that although it begins life as a female, it may metamorphose into a male later in life. Only a small percentage of female rock wrasse become males. Using its pectoral fins like oars, the wrasse swims in a distinctive looping motion. Sharp, projecting canine teeth, which protrude from a diminutive mouth, pick small invertebrates from seaweed growing on the rocks. Like the senorita (p.51), this fish dives into sand for protection. It remains there, with only its head exposed, until the danger has passed. Difficult to approach, the rock wrasse changes direction when followed. It lives up to fourteen years.

Male

Female

Juvenile; 1 1/2 inches long

California Sheephead

Semicossyphus pulcher

The sheephead is classified in the same family as the senorita (p.51) and rock wrasse (p.52) but is much larger. Beginning life as a female, it changes sex after about eight years of age, when the female sex organs metamorphose into male sex organs. Sex transformation takes less than a year. Although each male has his own territory, should he be eliminated, a resident female will quickly change into a male. The sheephead is a curious fish, and it is not unusual to find an adult tagging along behind while you explore the reef. The male sheephead may live up to fifty years.

Appearance: Male has a white chin, black head, red or pink middle, and black tail. Forehead is defined by a prominent bump. Female is pinkish with a white chin. Adults have imposing canine teeth. Juvenile is salmon-colored with black dots on fins and a black-and-white horizontal stripe across the sides of its body.

Size: Male to 3 feet; female to 2 feet; juvenile to 4 inches in length.

Where Found: Meandering around the rocky reefs and kelp bed.

Feeds On: Clams burrowed deep into sand, abalone, crabs, and lobsters.

Appearance: Distinctive pointed snout and unique forked tail. Color matches seaweed in the vicinity: yellow or mottled brown in kelp (pp.12,13), purple or reddish in red algae, such as braided hair (p.14), and green in surf grass (p.15). Juveniles are tan colored.

Size: To 2 feet in length.

Where Found: Around red, brown, and green marine plants.

Feeds On: Small crustaceans and small fishes such as kelp surfperch (p.45) and senorita (p.51).

Giant Kelpfish
Heterostichus rostratus

This unique-looking fish blends into its surroundings, and not solely by color. Orienting its leaflike body in the same direction as the seaweed, it sways in the surge, mimicking the motion of the plant fronds. Juveniles school around the giant kelp, until they reach 2 or 3 inches, when they develop adult colors and begin a solitary life. Seaweed teems with this common and dramatically beautiful fish. Be sure to seek out the giant kelpfish in its varied habitats. Discover for yourself its limitless diversity of colors and patterns. It is guaranteed to be worth your effort.

California Halibut

Paralicthys californicus

The halibut's large size separates it from other flatfish. Born with an eye on each side of its head, one eye slowly migrates to the other side. By adulthood, both eyes can be found only on one side of the head. In this species, slightly more than half are left-eyed. A fascinating swimmer to watch, this flatfish's ripply swimming motion is surprisingly effective in spiriting itself away in a hurry. Predators include sharks, such as the leopard shark (p.32), sea lions (p.58), harbor seals (p.59), and fishing enthusiasts.

Appearance: Large black spots adorn a flat, gray body. Both eyes usually on its left side. Mouth is filled with lots of sharp, pointed teeth.

Size: To 5 feet in length.

Where Found: Lying exposed on the sandy bottom, buried in the sand with just its eyes and mouth exposed, and swimming in an undulating motion over the sandy bottom or reef.

Feeds On: Fishes, squid, and sometimes octopuses.

Brown Pelican
Pelecanus occidentalis

Appearance: Huge, bulky, and prehistoric looking, with a long, flat bill and an immense throat pouch. Adult bird has a grayish-brown body and white head with feathers standing up in a spiky, punklike style. Breeding birds are chestnut on the back of the neck. Juvenile has a brown head and paler body.

Size: To 4-foot wingspan.

Where Found: Flying in "V" formation skimming just over the water, diving into the sea for prey and congregating on rocks by the Cove.

Feeds On: Fishes and squid.

The pelican flies so closely over the water, it almost grazes the surface with its wings. To become airborne, it "runs" on the water's surface while flapping its wings wildly. When sighting prey, the pelican plummets into the water from heights of up to 30 feet, grabs its catch, and returns to the surface to swallow. Under water, its pouch expands like a balloon to suck in small fish. Known to exist for at least 30 million years, this bird was threatened with extinction in the 1970s due to the chemical pesticide DDT. While it has made somewhat of a recovery, it remains on the endangered list.

Brandt's Cormorant

Phalacrocorax penicillatus

The cormorant can be seen warming itself on rocks with wings outstretched like solar panels. Its breeding area is found along the precariously narrow cliff ledges above the caves. The male first builds a nest of surf grass (p.15), then entices the female to lay eggs. The cormorant's underwater swimming method contrasts sharply with the pelican's dive-bomb method. Calmly sitting on the surface of the water, the cormorant effortlessly disappears under water using small stones often found in its gut as a kind of diver's weight belt. The cormorant has been seen to depths of over eighty feet.

Appearance: Black waterbird resembles a cross between a crow and an eel. Swimming under water, its head and neck extend forward and its wings tuck in tight against its body. Breeding birds manifest a cobalt-blue color on the throat pouch. Juvenile is dark brown with lighter under-parts.

Size: To 3 feet in length.

Where Found: Sitting on water, swimming under water, and perched in groups with its close relative, the brown pelican (p.56).

Feeds On: Fishes, squid, and crabs.

California Sea Lion
Zalophus californianus

Appearance: Tan to brown. Sleek looking with a narrow, pointed snout and little ears sticking out the sides of its head.

Size: To about 7 feet in length.

Where Found: Whizzing around the kelp bed and sunning itself on the nearby rocks.

Feeds On: Fishes, squid, and octopuses.

The sea lion's solid color, streamlined body, and little protruding ears readily distinguish it from a harbor seal (p.59). The sea lion is also nosy, playful, and fearless. One might zoom toward you at breakneck speed, stop directly in front of your face to stare, or follow behind to see where you are going. Unlike a human, a sea lion has no internal temperature regulation. Instead, it engages in a behavior called "rafting"— floating on the surface and holding its fins straight up in the air. By letting air or sun reach the network of tiny veins in its flippers, the blood is warmed or cooled. As a safety precaution, sea lions huddle together when rafting.

Harbor Seal

Phoca vitulina

The harbor seal's mottled color, earless head, and doe-eyed, plump appearance clearly differentiate it from the sea lion (p.58). It also averages 2 feet smaller and several hundred pounds less than the sea lion. Baby seals are as friendly and curious as sea lions, but adults are often wary and shy. The reserved harbor seal is also quiet compared to the gregarious sea lion's incessant bark. A seal often engages in a special sleeping behavior called "bottling"—snoozing bolt upright while bobbing in the water with just its nose above the surface.

Appearance: Gray to black with white spotting. Blubbery and blimplike in shape with no neck. Ears appear only as slits on the sides of its head.

Size: To 5 feet in length.

Where Found: Basking on the rocks nearby, people-watching with just its head exposed, and gracefully swimming through the kelp bed.

Feeds On: Fishes, and molluscs such as squid and octopuses.

Help to limit further cliff erosion and return the squirrels to their normal numbers. Please do not feed them!

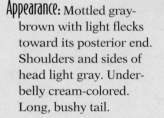

California Ground Squirrel
Spermophilus beecheyi

The ground squirrel is of great concern because feeding by well-meaning visitors has dramatically increased its numbers. Continual burrowing into the fragile cliffs further erodes the delicate ecosystem and forces habitat competition with the brown pelican (p.56) and Brandt's cormorant (p.57). The squirrel's natural enemies, which include large hawks, coyotes, eagles, foxes, wildcats, and badgers, no longer exist here. Without predators to keep the population in check and with an overabundant food supply, frequent litters are causing a population explosion. The gestation period for the California ground squirrel is about 30 days; litters average 7 young.

Appearance: Mottled gray-brown with light flecks toward its posterior end. Shoulders and sides of head light gray. Underbelly cream-colored. Long, bushy tail.

Size: To 18 inches in length (including tail).

Where Found: At street level where shrubs and trees give way to cliff areas.

Feeds On: Seeds, berries, plants, road kills, and most anything people toss its way.

Sightings Checklist

MARINE PLANTS

____ Braided Hair
____ Feather Boa Kelp
____ Giant Kelp
____ Surf Grass

INVERTEBRATES

____ Aggregating Anemone
____ California Spiny Lobster
____ Giant Keyhole Limpet
____ Green Abalone
____ Kellet's Whelk
____ Kelp Limpet
____ Knobby Sea Star
____ Mossy Chiton
____ Norris's Top Snail
____ Owl Limpet
____ Sand-Castle Worm
____ Striped Shore Crab
____ Sulfur Sponge
____ Sulfur Sponge Tylodina
____ Two-Spotted Octopus
____ Wavy Top Turban

VERTEBRATES-FISH

____ Bat Ray
____ Blacksmith
____ Black & Yellow Rockfish
____ Black Surfperch
____ Cabezon
____ California Halibut
____ California Moray Eel
____ California Sheephead
____ Garibaldi
____ Giant Kelpfish
____ Halfmoon
____ Kelp Bass
____ Kelp Surfperch
____ Leopard Shark
____ Opaleye
____ Pacific Sardine
____ Rock Wrasse
____ Round Stingray
____ Sargo
____ Senorita
____ Topsmelt
____ Walleye Surfperch
____ White Surfperch
____ Zebraperch

VERTEBRATES-BIRDS

____ Brandt's Cormorant
____ Brown Pelican

VERTEBRATES-MAMMALS

____ California Sea Lion
____ Common Ground Squirrel
____ Harbor Seal

Notes

Bibliography

Brandon, J.L. & F.J. Rokop (1985). *Life Between the Tides*.
American Southwest Publishing Company of San Diego, San Diego.

Connor, J. (1993). *Seashore Life on Rocky Coasts*. Monterey Bay
Aquarium Foundation, Monterey.

Dawson, E.Y. (1966). *Seashore Plants of Southern California*.
University of California Press, Berkeley and Los Angeles.

Goodson, G. (1988). *Fishes of the Pacific Coast*. Stanford
University Press, Stanford.

Gotshall, D.W. (1989). *Pacific Coast Inshore Fishes*. Third Edition.
Sea Challengers, Monterey.

Gotshall, D.W. & L.L. Laurent (1979). *Pacific Coast Subtidal
Marine Invertebrates*. Sea Challengers, Monterey.

Love, R.M. (1991). *Probably More Than You Want to Know About
the Fishes of the Pacific Coast*. Really Big Press, Santa Barbara.

Miller, J.M. & R.N. Lea (1972). *Guide to the Coastal Marine Fishes
of California*, Fish Bulletin No. 157. California Department of Fish
and Game, Sacramento.

Morris, R.H., D.P. Abbot & E.C. Haderlie (1980). *Intertidal
Invertebrates of California*. Stanford University Press, Stanford.

North, W.J. (1976). *Underwater California*. University of California
Press, Berkeley.

Schaelchlin, P.A. (1988). *La Jolla, The Story of a Community
1887-1987*. Friends of the La Jolla Library, La Jolla.

Index